POLISH CHICK FOR BEGINNERS

Essential Guide To Raising Cook, Ducks, - Best Practices, Housing, Nutrition, Health, And Profitable Techniques For Success

Holden bodhi

Contents

Overview Of Polish Poultry Production 9
- Comprehending Polish Chickens 9
- The Origins And History Of Polish Chickens 11
- The Advantages Of Polish Chicken Farming 13
- An Overview Of Fundamentals Of Farming..................15

CHAPTER TWO..19
- Knowing Polish Breeds Of Chicken 19
  - Overview Of Polish Chickens 19
  - Popular Breeds And Types 21
  - Qualities And Attributes 23
  - Selecting The Best Breed For Your Requirements 26

CHAPTER THRCHAPTER ONE 9EE
....29
- Establishing A Farm For Chickens 29
  - Selecting The Ideal Site 29
  - Crucial Tools And Materials................................34
  - Providing Appropriate Space And Ventilation 37

CHAPTER FOUR..41
- Constructing And Keeping Up With Chicken Coops 41
  - Cooperative Architecture And Building 41
  - Insulation And Ventilation 43
  - Frequent Upkeep And Sanitation 45

CHAPTER FIVE..49

- Well-Being And Health 49
  - Overview Of Avian Health 49
  - Typical Health Problems With Polish Chickens 51
  - Immunisations And Preventative Steps 52
  - Handling Illnesses And Parasites 54
- CHAPTER SIX 57
  - Polish Chicken Breeding 57
    - Overview Of Polish Chicken Breeding 57
    - Introduction To Poultry Breeding 60
    - Choosing Breeding Stock 62
    - Development And Raising Of Chicks 65
- CHAPTER SEVEN 69
  - Everyday Handling And Administration 69
    - Overview Of Daily Maintenance 69
    - Observing The Behaviour Of Chickens 74
    - Maintaining And Managing Records 76
- CHAPTER EIGHT 79
  - Production And Management Of Eggs 79
    - Knowing The Cycles Of Egg Production 79
    - Gathering And Preserving Eggs 81
    - Handling Typical Egg Problems 82
- CHAPTER NINE 87
  - Overview Of Seasonal Factors 87
    - Changing Up Care For Various Seasons 87

Handling The Stress Of Heat And Cold 90
Seasonal Wellness Advice 92
CHAPTER TEN 95
Promoting And Getting Rid Of Your Chickens 95
Recognising Demand In The Market 95
determining prices and distribution channels 96
how to market your poultry enterprise 98
Recognising Demand In The Market 100
Determining Prices And Distribution Channels 102
How To Market Your Poultry Enterprise 105
CHAPTER ELEVEN 107
Faqs & Troubleshooting 107
Solving Typical Problems In Polish Chicken Farming 107
Commonly Asked Questions 109
Sources Of Additional Support 111
Typical Problems And Their Fixes 113

Copyright © 2024 by Holden bodhi

All rights reserved.

No part of this publication may be reproduced, distributed, or transmitted in any form or by any means, including photocopying, recording, or other electronic or mechanical methods, without the prior written permission of the publisher, except in the case of brief quotations embodied in critical reviews and certain other non commercial uses permitted by copyright law.

DISCLAIMER

The information provided in this book, is intended for educational and informational purposes only. The content is based on research, personal experiences, and general knowledge about farming. It is not intended to substitute professional advice or expert consultation. Readers are encouraged to seek professional guidance when implementing any practices or techniques discussed in this book.

The author and publisher make no representations or warranties of any kind regarding the accuracy, applicability, or completeness of the contents of this book. Any reliance you place on such information is strictly at your own risk. The author and publisher shall not be held liable for any damages, losses, or injuries resulting from the use of the information provided.

Additionally, the author does not endorse, recommend, or affiliate with any individual, product, service, website, organization, or brand mentioned or referenced in this book. Any such references are solely for informational purposes, and no warranty or guarantee is implied. The inclusion of these references does not imply any endorsement or partnership by the author.

By reading this book, you acknowledge and accept that the author and publisher are not responsible for any consequences arising from your use of the information pro

CHAPTER ONE

Overview Of Polish Poultry Production

Comprehending Polish Chickens

Poultry fans will find Polish chickens to be an interesting choice because of their unique appearance and endearing characteristics. They are beneficial both aesthetically and practically because of their distinctive feather crests and placid disposition. An overview of what makes Polish chickens unique and why they are a desirable addition to any farm can be found in this section.

Important Features

Polish chickens are distinguished by their flamboyant head crests, which come in a variety of designs and hues. Because of their peaceful disposition, these birds are appropriate for both beginning and seasoned farmers. They are frequently a favourite among both adults and children because to their amiable disposition.

Why Opt for Polish Chickens

Adding Polish chickens to your collection of fowl will give it a distinctive look. In many backyard flocks, their attractive look is the focal point, and their amiable nature can make them excellent companions. They can also bring a touch of elegance to your farm and act as conversation starters.

Beginning Your Polish Chicken Journey

It's important to comprehend Polish chickens' fundamental needs and requirements before getting into the specifics of rearing them. This introduction will walk you through the first stages of establishing a Polish chicken farm, from choosing the appropriate breed to comprehending the requirements for their upkeep.

Having Reasonable Objectives

Setting specific objectives for your Polish chicken farming endeavour can help you get started in the right direction. Having clear goals will help you make better decisions while you farm, whether your goal is to raise them for eggs, show, or just as pets.

The Origins And History Of Polish Chickens

Where Polish Chickens Came From

Polish hens have a colourful and varied history that dates back to Eastern Europe. These hens are native to Poland and have long been raised for poultry in the area. This section explores their historical past, including background information on their evolution and importance.

Early Records and Breeding

Polish hens were first mentioned in 16th-century historical documents and pictures, where their unique appearance was mentioned. Their distinctive traits are a result of their breeding history being entwined with the growth of other decorative poultry breeds.

Cultural Importance

Polish chickens have been valued in many different civilisations throughout history. They were praised for their beauty and distinctive feather crests and were displayed in local fairs and shows. Their allure and charm are enhanced by their cultural significance.

Contemporary Development

Beyond their home country of Poland, Polish chickens have become more and more popular in recent years. The unique characteristics of the breed have been improved and preserved by breeders all over the world. This section looks at the evolution of Polish chickens and the ongoing rise in popularity.

Preservation and Conservation

In order to guarantee that future generations will appreciate Polish chickens' distinctive qualities, continuing conservation and preservation efforts are being made. The goal of organisations and breeders is to uphold the standards of the breed and encourage its continued existence.

The Advantages Of Polish Chicken Farming

Appeal to the Senses

Polish hens are visually appealing, which is one of the main advantages of keeping them. They are very stunning with their prominent head crests and diverse colours of plumage. This section explains how their distinct appearance can improve the way your collection of fowl looks as a whole.

Relationships and Personality

Polish chickens are renowned for having a laid-back and amiable disposition. They are simple to maintain and frequently end up as cherished pets. This section explains how their benign character makes them appropriate for both individuals and families.

Production of Eggs

Polish hens do lay a fair amount of eggs, despite not being kept primarily for their eggs. In general, their eggs are high-quality and white. This section discusses their ability to lay eggs and how it stacks up against other breeds.

Flexibility in a Range of Situations

Polish hens can live in small coops or on huge farms, depending on the situation. This section examines how their adaptability might help with diverse poultry setups and how they can flourish under distinct circumstances.

Relevance to Education

Both adults and children can learn from the experience of raising Polish chickens. Understanding their history, care, and behaviour can bring you important insights into animal husbandry and poultry farming. The advantages of raising Polish hens for schooling are highlighted in this section.

An Overview Of Fundamentals Of Farming

Establishing the Cooperation

The first step in creating a good environment for Polish hens is building a sturdy coop. The basic characteristics of a chicken coop are described in this part, along with the amount of space needed, ventilation, and predator protection.

Nutrition and Feeding

For Polish hens to be healthy and happy, they must be fed properly. To achieve the best possible growth and egg production, this section offers advice on feeding procedures, including what kinds of feed, supplements, and feeding schedules to use.

Well-being and Upkeep

Maintaining Polish hens in good health requires routine health exams and care. Common health problems, preventative methods, and the significance of routine veterinary care are all covered in this section.

Producing and Bringing Up Chicks

It's crucial to comprehend the fundamentals of breeding and chick care if you intend to breed Polish chickens. An overview of breeding procedures, incubation, and raising healthy chicks from hatch to maturity is given in this section.

Maintaining and Managing Records

Maintaining thorough records of your Polish chicken farm's upkeep, productivity, and health is essential to its successful operation. This section highlights the need of maintaining records and offers pointers for effectively managing and monitoring your flock.

CHAPTER TWO

Knowing Polish Breeds Of Chicken

Overview Of Polish Chickens

Polish hens are a special breed distinguished by their lovely personalities and unusual appearance. These hens are Polish natives who have contributed centuries to the history of poultry. Their remarkable feather crests on their heads, which give them a unique, almost humorous appearance, are an easy way to identify them. The development and history of this breed offer a fascinating context for comprehending its place in contemporary chicken production.

Origin and History

Polish chickens have been around since the 16th century, when their egg output and decorative value made them highly valued. Although they were probably derived from many breeds in the area, their name comes from their Polish heritage. Polish chickens have developed throughout the ages into a breed that is useful for a variety of farming applications in addition to being aesthetically beautiful.

Special Qualities and Look

Polish hens are known for having beautiful feather crests that come in a variety of sizes and shapes. Their unique look is mostly due to the crests that are the product of selective breeding. Polish chickens are available in a variety of hues and designs, such as white, black, or a combination of the two. Their bodies and necks are also covered in feathers, which enhances their ornamental attractiveness.

Goal and Role

Polish hens are preserved for more than just aesthetics; they have useful uses as well. They are good medium-sized egg layers with a peaceful temperament that makes them ideal for backyard flocks. Still, in many farming circumstances, their ornamental value often prevails.

Popular Breeds And Types
The Most Well-liked Polish Breeds

Polish chickens are divided into a number of breeds, each having unique traits. The White Crested Black Polish, Golden Polish, and Silver Polish are the most popular types. Every breed is different from the others and fits well in a certain farming environment.

Polish with White Crested Black Hair

The stunning black feathers and noticeable white crest of the White Crested Black Polish are its most notable features. This breed is well-liked among hen aficionados because it is not only gorgeous but also quite resilient. Its durability and graceful beauty add to its appeal.

gilded polish

The broad, rounded crest and golden feathers of the Golden Polish set it apart. Because of this breed's colourful look and environment adaptability, it is frequently chosen. It is a favourite in poultry fairs and exhibits because of its exquisite plumage.

Silver Glaze

The Silver Polish's striking crest is complemented by its elegant, silvery-white feathering. This breed is renowned for both its good egg-laying skills and serene demeanour. The Silver Polish is an adaptable breed that does well in a variety of settings and is frequently chosen because to its pleasing appearance.

The Applications of Breed Variations

There are variations in colour and feather pattern within these primary breeds. These differences may have an impact on the breed's general look and suitability for various farming applications. Comprehending these variances aids farmers in choosing the ideal breed for their particular objectives, be it egg production or aesthetic appeal.

Qualities And Attributes
Physical Characteristics

Large, feathered crests and characteristic feather patterns are among the unusual physical characteristics of Polish chickens that make them well-known. These characteristics, which come from selective breeding, add to their allure. Although their crests might occasionally obstruct their vision, Polish hens generally adjust well to their surroundings.

Feather Crests

Perhaps the most recognisable feature of Polish chickens is their feather crest. It might be as small as a modest tuft or as big as a massive headpiece that covers the face almost entirely. Not only is this feature decorative, but it also serves to shield their head from the weather.

Anatomy

Polish hens are built with a compact, balanced body. They usually have powerful legs and a wide chest, which contribute to their general functionality and health.

Though they have decorative traits, they are hardy birds that can survive in a variety of environments.

Behaviour and Temperament

Polish hens are renowned for being kind and amiable. They are good for families and backyard flocks since they are typically peaceful and simple to manage. They are less likely to become violent or too territorial due to their placid disposition.

Social Conduct

These hens are gregarious creatures who take pleasure in other birds' companionship. They can adjust to different flock dynamics and generally get along well with diverse breeds.

Their amiable disposition also permeates their contact with people, as they frequently grow used to being handled and even form close relationships with those who look after them.

Maintenance Needs

Polish chickens have unique demands that must be met, such as maintaining their crests and making sure they have enough shelter.

Their feathered crests can harbour pests, so it's vital to regularly inspect for mites and other parasites. To avoid overgrowth, their crests may also need to be trimmed on a regular basis.

Selecting The Best Breed For Your Requirements

Evaluating Your Farming Objectives

Prior to choosing a breed of Polish chicken, you should consider your agricultural objectives. Do you need hens to produce eggs, or are beautiful birds more your style? Knowing your main goals will help you select the breed that best meets your requirements.

Functional versus Ornamental

Because of their eye-catching looks, breeds like Silver Polish or Golden Polish may be perfect if you are primarily concerned with your hens' aesthetic appeal. However, if you also need your chickens to produce eggs well, you may want to look for breeds that strike a balance between appearance and output.

Environment and Climate

Take into account the weather and surroundings where your hens will be housed. While certain Polish breeds do better in milder regions, some are better suited to colder ones. Making sure your breed is acclimated to the climate where it lives will improve its general health and output.

Realistic Aspects

Practical considerations including space, food availability, and maintenance needs should be made when selecting a breed of Polish chicken. Make sure you have the means and dedication required to meet the unique requirements of the breed you select.

Room & Accommodation

For survival, Polish hens need enough room and protection. Their crests can make them more vulnerable to environmental changes, so it's important to give them a safe, well-ventilated coop. Their general health and happiness will be enhanced if your house satisfies their needs.

Costs and Budget

Think about the expenses involved in purchasing and maintaining Polish hens. Even while a breed's initial cost of purchase may differ, you need also consider continuing costs such food, bedding, and veterinarian care when making your choice. A profitable and long-lasting agricultural experience can be ensured by setting aside money for these expenses.

CHAPTER THREE

Establishing A Farm For Chickens

Selecting The Ideal Site

The productivity and well-being of your Polish hens depend greatly on where you decide to set up your farm. The location should ideally have sufficient drainage and be raised to prevent flooding. Because hens are susceptible to a number of diseases if their surroundings is too damp, make sure the area is well-drained. Select a spot where your flock will be naturally shaded by trees or plants and protected from the elements. This will help keep your flock comfortable in the summer and protect it from chilly winter winds.

Organising the Layout of Your Farm

An intelligent design improves comfort and efficiency. Make a map of your farm's many zones first, including the run area, feed storage, chicken coop, and access routes. Make sure the coop is positioned to reduce exposure to direct sunshine and high winds. Think about how daily tasks are completed, such as cleaning, feeding, and collecting eggs. A rational layout lowers labour costs and enhances administration in general.

Legal Aspects and Licenses

Do your homework on local zoning laws and restrictions pertaining to chicken farming before you begin. Permits are needed in many places, and there are rules about farm structures, waste disposal, and noise levels.

Make sure you comply to prevent penalties or legal problems later on. For comprehensive information on necessary permissions and regulations, get in touch with the municipal planning department or agricultural extension office in your area.

Getting the Land Ready

Debris removal and making sure the earth is ready for building are two aspects of land preparation. The coop and run site should be levelled, and any large rocks or tree stumps should be removed. Consider adding compost or organic materials to the area if the soil quality is low. The lifespan of your constructions is ensured by proper site preparation, which also provides a strong basis for construction.

Financial Planning and Budgeting

An upfront financial outlay is necessary for building, equipment, and other necessities when starting a chicken farm. Make a thorough budget that accounts for labour, supplies, and any unforeseen costs. Include recurring expenses for things like feed, vet care, and utilities. Maintaining the sustainability of your farm and controlling expenses are made easier with a well-defined financial plan.

Creating the Run and Coop

Comprehending the Needs for a Chicken Coop

Like any poultry, Polish chickens need a coop that is useful, comfortable, and safe. Each chicken should have enough room in the coop; generally speaking, 2-4 square feet should be allotted to each chicken. Nesting boxes for laying eggs, roosting bars for sleeping, and simple access for upkeep and cleaning should all be included in the design. Make sure the coop is safe from predators by building it with sturdy materials and appropriate methods.

Creating the Run Area Design

There should be plenty of room in the run area for the hens to roam around and forage. Eight to ten square feet are the ideal size for each bird. Provide elements that will keep hens occupied and healthy, such as access to fresh grass, dust baths, and sheltered spots. Sturdy fencing is needed to enclose the run and keep the chickens from escaping as well as predators out. To keep predators from burrowing, make sure the fencing is buried several inches below the surface.

Providing Defence Against Predators

Protecting your Polish hens against predators requires predation-proofing. Dogs, foxes, hawks, and raccoons are examples of common predators. Use hardware cloth or welded wire to reinforce the coop and run rather than chicken wire, which is prone to breaking. Keep an eye out for any possible weak points and secure all openings. A predator deterrent system like as motion-activated lights or sirens can also be included.

Including Lighting and Ventilation

In order to keep everyone comfortable and avoid respiratory problems, the coop must have adequate ventilation. Create vents or windows that you may open or close based on the weather. The health and yield of eggs from the hens also depend on natural light. Windows should be positioned to let in sunlight, however direct sunlight can cause the coop to overheat.

Formulating a Plan for Cleaning and Maintenance

A well-thought-out coop and run should make cleaning and maintenance simple. Use a slanted floor or detachable dropping trays to funnel garbage into a collection location. Provide access points that make it simple to remove waste and bedding. Frequent cleaning promotes the general health of your flock by preventing the growth of dangerous bacteria and parasites.

Crucial Tools And Materials

Choosing the Proper Watering and Feeding Systems

Your Polish hens' health depends on whether you feed them high-quality feed and give them clean water. Select a waterer and feeder that are both easy to clean and can hold the number of hens you own. Systems that are gravity-fed or automatic can help to maintain a steady supply of food and water by minimising the need for frequent refills.

Selecting Materials for Bedding

Bedding contributes to the coop's ability to stay tidy and cosy. Sand, wood shavings, hay, and straw are among the options. Each variety has advantages of its own: wood shavings and sand are easier to clean, but straw and hay are wonderful for insulation. To keep an area smelling fresh and preventing odours, change the bedding on a regular basis.

Purchasing Safety and Health Supplies

Supplies for health and safety are necessary to oversee the welfare of your herd. Ensure you have an ample supply of basic veterinary products, including dewormers, antibiotics, and first aid items. In order to avoid illness, make a strategy for routine physical examinations and immunisations. Use hand sanitisers and foot baths as biosecurity precautions to lessen the chance of illness transmission to your flock.

Instruments for Upkeep of the Cooperative

Maintaining the coop and run in good shape is ensured by having the appropriate tools. A shovel for changing bedding, a rake for cleaning the run, and basic repair equipment like screwdrivers and hammers are essential. Frequent maintenance keeps problems from getting worse and prolongs the life of your structures and equipment.

Surveillance Tools

You may help keep the coop's conditions at their ideal level by keeping an eye on monitoring tools like humidity and temperature gauges. Install hygrometers and thermometers to monitor environmental conditions and make necessary adjustments. This gadget ensures your hens' comfort and well-being by mitigating problems caused by high or low humidity levels.

Providing Appropriate Space And Ventilation

The Significance of Sufficient Airflow

In order to keep the environment within the coop healthy, proper ventilation is essential. It lowers the risk of respiratory disorders and other health problems by assisting in the control of temperature, humidity, and air quality. Provide windows or vents that may be adjusted to provide enough airflow without creating drafts when designing the coop.

Space Efficiency in Design

Every chicken has adequate space to move about comfortably when there is efficient utilisation of available area. In order to avoid tension and hostility among the flock, the coop should have adequate space. Additionally, the run space needs to be big enough to meet the hens' needs for foraging and exercise.

Controlling the Humidity and Temperature

Controlling humidity and temperature is crucial for the wellbeing of chickens. To keep the coop cool in the summer and warm in the winter, make sure it is insulated. To control ventilation and lower humidity, use fans or ventilation systems. Keep an eye on everything and make any adjustments to keep your flock comfortable.

Establishing a Cosy Home Environment

In addition to enough room and ventilation, a comfortable setting also include elements that support wellbeing. Provide places for dust baths, roosting bars, and nesting boxes. To keep the chickens from overheating, make sure the coop is tidy and draft-free. You should also provide shade for the run area.

Keeping an eye on things and making adjustments

Keep an eye on the coop and run's conditions to make sure they continue to be ideal. Examine the hens for any indications of discomfort or health problems, and make any necessary adjustments to the temperature, ventilation, and space. Maintaining a healthy and prolific floc requires paying attention to these aspects.

CHAPTER FOUR

Constructing And Keeping Up With Chicken Coops

Cooperative Architecture And Building

To safeguard the health and welfare of your flock, meticulous planning is necessary when designing and building a chicken coop. A sturdy coop gives your hens comfortable living quarters as well as protection from the weather and predators.

Size and Layout: Take into account the amount of hens you intend to maintain when planning your coop. As a general guideline, each chicken should have 2 to 4 square feet of room within the coop and at least 10 square feet outside in the run. Nesting boxes, roosting bars, and sufficient floor space should all be included in the coop. Make sure the hens have adequate space to go around freely and comfortably.

Materials and Construction: Use strong, weather-resistant materials. Commonly used materials for building coops are plastic, metal and wood. Wood is well-liked for its ease of construction and insulating qualities, although rot requires treatment. Although metal coops are strong, they may not be as insulated. Although they are pest-resistant and simple to maintain, plastic coops might not offer as much insulation.

Design Elements Include elements that make maintenance and access simple. A well-thought-out coop ought to have:

- Nesting Boxes: Ensure that nesting boxes are easily accessible for collecting eggs by placing them approximately 12 inches above the ground.

- Bars for Roosting: Place bars for roosting at least eighteen inches above the ground. For safe sleep, hens like to roost high.

- Access Doors: Provide doors that make it simple to clean and maintain the coop. Make sure that these doors are locked to keep predators out.

- Ventilation: Keeping the coop's inside air clean and humidity levels low requires proper ventilation. Put high-up vents on the roof or walls to let out moisture and hot air.

Insulation And Ventilation

To keep the coop's atmosphere healthy, proper insulation and ventilation are crucial. They lessen the chance of respiratory issues in hens and assist with controlling temperature and humidity.

Proper ventilation keeps the coop's interior cool, humid, and ammonia-free while ensuring a steady supply of fresh air. Put vents in the vicinity of the roofline to let warm air out. Make use of windows or vents that may be adjusted to open or close in response to the weather.

Vents shouldn't be placed exactly across from one another as this can lead to drafts.

Insulation Keeping the coop warm in the winter and cool in the summer is made possible by insulation. Foam boards and fibreglass are two types of insulation that can be put in the walls and ceiling. Make sure insulation is shielded from insects and dampness. If it's chilly outside, think about utilising an electric heater or a heat lamp in colder climates to stay warm.

Seasonal Adjustments: Depending on the weather, modify the insulation and ventilation. Reduce ventilation in cooler months to keep heat in while making sure there's still adequate airflow to avoid condensation. To lower the temperature inside the coop during the warmer months, make the most of the ventilation and think about providing shade.

Keeping the Air Quality Good Keep an eye out for indicators of poor air quality, such as high dust, an ammonia smell, or respiratory problems in your hens. If you need to increase ventilation and lower humidity, install a fan. Regular cleaning will help to keep waste and mould from building up in the coop.

Frequent Upkeep And Sanitation

The health of your hens and the longevity of the building depend on you keeping the coop clean and functional. In addition to ensuring a comfortable living environment, routine cleaning and maintenance assist prevent disease and control pests.

Daily Tasks: Make sure the coop is safe, check the food and water levels, and clean up any droppings from the floor and roosting bars. Examine the coop for indications of wear or damage, and take quick action to fix any problems.

Weekly Cleaning: Give the coop a deeper cleaning once a week. After removing all of the bedding, use a light disinfectant to clean the walls and floors, and then replace with brand-new bedding. Clear any collected debris from the roosting bars and nesting boxes. To make sure the ventilation system is operating correctly, inspect and clean it.

Monthly Maintenance: Checking for symptoms of disease or pests, fixing any problems with the walls or roof, and examining the coop for structural damage are all part of the monthly maintenance routine. Make sure all of the doors and locks are operating properly, and fix anything that needs fixing.

Seasonal Maintenance Conduct a thorough cleaning and examination as the seasons change. Check insulation and plug any gaps in the coop's construction to get it ready for winter in the autumn. Make sure the coop is ready for warmer weather in the spring by inspecting the ventilation and fixing any damage caused by the winter.

Pest Control: Frequently look for indications of pests like rats, lice, or mites. To keep pests away, keep the coop dry and clean. If necessary, use natural deterrents or traps. Make sure every food and water container is hygienic and doesn't draw in bugs.

Keeping your hens healthy and productive is ensured by keeping their coop neat and organised, which adds to a profitable and pleasurable experience of raising chickens.

CHAPTER FIVE

Well-Being And Health

Overview Of Avian Health

Polish chickens' general well-being and productivity depend on their general health and wellness. Polish chickens are distinguished by their distinctive feathered crests and amiable disposition, but much like other poultry, they are prone to a number of health problems. A combination of proper husbandry practices, preventative measures, and prompt action when issues develop are necessary to ensure their health. The vital components of maintaining the health and well-being of your Polish hens are covered in this section.

The Significance of Balanced Diet

Poultry health is mostly dependent on proper diet. Polish hens need a diet that is well-balanced and full of important elements, such as vitamins, minerals, and proteins. Feeding should be based on the animal's size, age, and intended use (e.g., producing meat or eggs). Strong bones, a healthy immune system, and healthy feather growth are all maintained with the aid of high-quality poultry feed. Always have access to fresh water to maintain healthy digestion and general well-being.

Housing and Environmental Conditions

Polish chickens' living conditions have a big influence on their health. Maintaining a clean, well-ventilated coop is crucial for comfort and to avoid respiratory problems.

In order to prevent stress and the spread of illnesses, there must be enough space. Effective waste management, appropriate bedding, and routine coop cleaning are essential for avoiding the growth of parasites and dangerous pathogens.

Typical Health Problems With Polish Chickens

Issues with the Respiratory System

Because of their distinctive feather crests, which can impede their vision and make breathing difficult, Polish hens are more likely to experience respiratory problems. Colds, coughs, and more serious ailments like bronchitis and pneumonia are common respiratory issues. Coughing, sneezing, nasal discharge, and difficulty breathing are possible symptoms. These problems can be managed with frequent observation and timely treatment with the right drugs.

Skin Conditions and Feathers

In Polish hens, skin disorders and feather loss are rather prevalent. Skin irritation, feather loss, and itching can be brought on by lice, mites, and fungal infections. Early diagnosis and treatment of these illnesses are essential to avoiding more consequences. Good cleanliness habits and routine examination of the feathers and skin can aid in the early discovery and treatment of these problems.

Gastrointestinal Conditions

Polish hens are susceptible to digestive diseases such as crop impaction or enteritis. Ingestion of foreign items, insufficient water intake, or poor diet can all lead to these problems. A swelling crop, altered dropping patterns, or decreased appetite are possible symptoms. Numerous digestive problems can be avoided by offering a balanced diet and guaranteeing access to clean water. Veterinarian intervention may be necessary in severe situations.

Immunisations And Preventative Steps
The Vaccination's Function

For Polish hens, vaccination is an essential part of preventive healthcare. Vaccines guard against major diseases of chicken, including infectious bronchitis, Marek's disease, and Newcastle disease. Maintaining a regular immunisation schedule lowers the chance of disease outbreaks and aids in immunity development. To create a vaccination schedule that is right for your flock, speak with a vet.

Measures for Biosecurity

Strong biosecurity protocols must be put in place to stop the entry and spread of pathogens. This include keeping the chicken coop tidy, restricting access, and using appropriate hygiene. Early diagnosis and control of potential health concerns can be aided by routinely cleaning equipment and keeping an eye out for symptoms.

Continual Health Examinations

Polish hens need regular health tests to be kept in good condition. These examinations have to comprise keeping an eye out for symptoms, evaluating general health, and making sure that all medical procedures are being followed. Treatment and early intervention can stop minor health problems from getting worse before they become more serious.

Handling Illnesses And Parasites
Recognising Typical Parasites

Like any poultry, Polish chickens can have worms, lice, and mites, among other parasites. It is necessary to regularly examine feathers, skin, and droppings in order to identify these parasites. Itching, losing feathers, and losing weight are examples of symptoms that may point to a parasite infestation. Maintaining a clean environment and applying authorised antiparasitic medicines are two aspects of proper maintenance.

Putting Effective Treatment Plans Into Practice

Medication and environmental control are used in the treatment of parasites. In addition to giving antiparasitic drugs in accordance with veterinarian recommendations, the chicken coop needs to be properly cleaned and disinfected in order to get rid of any leftover parasites. It could be essential to conduct follow-up therapies and routine monitoring to guarantee total eradication.

Preventive Techniques

Maintaining sufficient diet, clean bedding, and good flock hygiene are all necessary to prevent parasite infections. The prevention of parasite transmission can also be aided by routine health examinations and timely treatment of any health problems. You may improve your Polish hens' general health by learning about common parasites and how to treat them.

This thorough manual addresses all the important areas of health and well-being for

Polish chickens, such as vaccination schedules, common ailments, parasite control, and preventative measures. You may safeguard your flock's welfare and encourage their long-term health and productivity by putting these strategies into effect.

CHAPTER SIX

Polish Chicken Breeding

Overview Of Polish Chicken Breeding

Polish hens have special traits and needs that must be understood in order to breed them. Polish hens are recognised for their characteristic crests and amiable disposition. Understanding the characteristics of the breed and how to pass them on to the following generation is the first step towards successful breeding. The basics of producing Polish chickens, such as their genetics, health issues, and breeding objectives, will be covered in this chapter.

Goals of Breeding and Genetics

A thorough understanding of Polish chicken genetics is essential for successful breeding. The breed can be selectively bred to improve desired qualities, as it has multiple colour variations and crest forms. The fundamentals of chicken genetics, including dominant and recessive traits and breeding objective setting, will be covered in this part.

It will also go over how to design breeding programs to reach particular goals, including more productivity or better-quality feathers.

Health Issues with Breeding

Successful breeding operations depend on having healthy breeding stock. Polish chickens are sturdy in general, but because of their distinctive crests, which can occasionally impair their vision, they may be more susceptible to certain health problems.

This section will cover frequent health issues in Polish hens, how to maintain a healthy breeding environment, and preventive measures. Immunisations, managing diseases, and the significance of routine health examinations will all be covered.

Methods and Practices of Breeding

Polish chickens of superior grade are primarily produced through efficient breeding methods. The practical side of breeding will be covered in this section, including how to pair chickens according to their qualities, comprehend mating behaviours, and take good care of the hens and roosters. In order to preserve genetic diversity, it will also provide advice on controlling breeding cycles and avoiding inbreeding.

Assessing the Success of Breeding

It's critical to assess the performance of your breeding program after breeding in order to determine what worked and what still needs improvement. This part will cover techniques for analysing the outcomes of your breeding endeavours, such as checking the qualities and health of the chicks, keeping an eye on growth rates, and modifying your breeding plan in response to the results.

Introduction To Poultry Breeding
A Brief Overview of Poultry Breeding

In the specialised field of poultry breeding, birds are chosen and mated in order to create offspring with desirable traits. An outline of chicken breeding, including its goals, advantages, and fundamental ideas that direct the procedure, will be given in this part. You may build a strong basis for growing Polish chickens or any other fowl by being aware of these fundamentals.

Genetic Concepts in the Breeding of Poultry

Breeding chickens heavily relies on genetics. Basic genetic concepts, including inheritance patterns, gene expression, and the transmission of genetic features from parents to children, will be covered in this part. It will go over ideas like sex-linked features, dominant and recessive genes, and how to use this information to make wise breeding decisions.

Choosing Breeding Stock

Breeding objectives must be met by carefully choosing the breeding stock. The selection criteria for healthy, superior birds—which include physical attributes, behaviour, and genetic heritage—will be delineated in this section. It will also include advice on how to evaluate prospective breeders' appropriateness and the significance of preserving genetic diversity.

Breeding Methods and Plans

To accomplish particular objectives in chicken breeding, a variety of methods and approaches can be employed. Different breeding techniques, including backcrossing, crossbreeding, and line breeding, will be covered in this section. The practical parts of putting these strategies into practice will also be covered, such as controlling breeding cycles, keeping records, and assessing breeding outcomes.

The Welfare and Handling of Reproductive Fowl

Successful breeding operations depend on the health and welfare of the breeding fowl. This section will discuss management techniques that enhance general well-being, preventive measures, and common health problems in poultry. Disease prevention, housing, sanitation, and nutrition will all be covered.

Choosing Breeding Stock
Overview of Choosing Breeding Stock

One of the most important factors in effective chicken breeding is choosing the appropriate breeding stock. The health, productivity, and characteristics of your offspring will be directly impacted by the calibre of your breeding birds. This section will offer a thorough guidance with important criteria and factors for choosing breeding stock.

Assessing Outer Features

Important predictors of breeding potential include physical characteristics. How to evaluate physical characteristics like body size, feather quality, and general appearance is covered in this section. It will also contain pointers on recognising desirable characteristics unique to Polish chickens and how these characteristics affect the calibre of progeny.

Evaluating Behavioural Characteristics

The general effectiveness of breeding stock can be influenced by behavioural features. The evaluation of behavioural features, including temperament, mating habits, and social interactions, will be covered in this section. It will also offer advice on how to select birds that display desired behaviours and how these qualities affect the results of breeding.

Genetic Origins and Pedigree

Making educated breeding decisions requires having a thorough understanding of the genetic heritage and ancestry of your breeding stock. This section explains the significance of pedigree records and genetic testing, as well as how to investigate and assess the genetic background of prospective breeders. Additionally, it will discuss how to apply this knowledge to preserve genetic variety and prevent inbreeding.

The well-being and state of the breeding stock

For breeding to be successful, your breeding stock's health and condition are essential. This section will describe how to evaluate a prospective breeder's general health, including how to look for physical anomalies, parasites, and disease symptoms. It will also include advice on how to take good care of and manage breeding birds to preserve their health and wellbeing.

Development And Raising Of Chicks
Overview of Chick Rearing and Incubation

The crucial phases of chicken breeding, incubation and chick rearing, call for meticulous control to guarantee the healthy development of the young. An outline of the incubation procedure and essential techniques for rearing healthy chicks from hatch to maturity are given in this section.

Methods of Incubation

Establishing the ideal environment for the development of eggs is essential to successful incubation. This section will discuss various incubation techniques, such as artificial incubation in incubators and natural brooding by chickens. It will include information on regulating humidity and temperature, rotating eggs, and keeping an eye on development to guarantee ideal circumstances for hatching.

Getting Ready for Chick to Come

Getting ready for the chicks' arrival is essential to ensuring their survival and immediate care. The setup of a brooder, including bedding, feeding supplies, and temperature control, will be covered in this section. It will also go over how to get ready for things like adjusting to the requirements of various breeds of chicks.

Practices for Chick Rearing

Healthy chicks need to be raised with the right supervision and care. Important guidelines for raising chicks, such as feeding, watering, and housing, are provided in this section. It will include instructions on how to check health and development, create an atmosphere that is conducive to growth, and move chicks to outside housing as they get older.

The Welfare and Handling of Developing Foetuses

Long-term success of rearing chicks depends on maintaining their health. This section will discuss common health concerns in chicks, including limb diseases and respiratory troubles, and offer advice on how to prevent and treat these conditions. It will also cover diet, immunisation regimens, and continuing care to guarantee healthy development and growth.

Assessing the Growth and Achievement of Chicks

Assessing the growth and well-being of your chicks aids in determining how well your incubation and rearing techniques are working. The techniques for assessing the behaviour, health, and growth of chicks are covered in this section. It will also cover how to prepare chicks for their future responsibilities in your flock and make adjustments to improve results.

Every segment is intended to offer a thorough manual for raising Polish chickens, covering everything from comprehending the fundamentals of poultry breeding to the particular needs for incubation and raising young birds. Please let me know if you require any more information or changes.

CHAPTER SEVEN

Everyday Handling And Administration

Overview Of Daily Maintenance

Maintaining your Polish hens' health and productivity requires daily attention and control. You can make sure your hens stay content, safe, and productive by implementing a schedule. This section offers a thorough overview of the everyday chores that are necessary for the management of chickens.

Watering and Feeding

Making sure your hens have clean water and fresh food is one of your most critical daily tasks. Polish chickens have different nutritional needs according on their age, weight, and stage of production, so feed should be tailored to meet those needs. Make sure water bottles and feed dispensers are always clean and full.

Replace any water or feed that has gone bad to avoid health problems.

Sanitation and Cleaning

It is imperative to conduct routine cleaning of the chicken coop and its environs to avert the accumulation of refuse and possible diseases. Cleaning feed and water containers, getting rid of dirty bedding, and checking the coop for mould or mildew growth are all daily chores. Make sure the coop has adequate ventilation to lower the humidity and ammonia levels, which can be harmful to the health of the chickens.

Examinations of Health

Every day, visually inspect each bird to look for indications of disease or damage. Keep an eye out any behavioural, feather, and physical changes. Illness is frequently indicated by sleepiness, irregular stools, or changes in appetite. As soon as possible, resolve any problems by speaking with a veterinarian or poultry specialist.

Safety and Security

Make sure the coop is protected from foxes, raccoons, and neighbourhood dogs, among other possible predators. Verify that all windows, doors, and fencing are secure and locked. Evaluate the general safety of the coop and make the required repairs to avoid mishaps or escapes.

Maintaining Records

Keeping thorough records of your daily care actions can help you monitor your hens' productivity and overall health. Keep track of feed intake, water use,

cleaning routines, and any health problems that are noticed. Making educated management decisions and spotting trends or issues requires the use of this data.

Normal Care Activities

Feeding Schedule

A regular feeding routine is essential to guaranteeing your Polish chickens get the right nourishment. Chickens are normally fed twice a day, though this might vary depending on the age and stage of production. Keep an eye on their feed intake to make sure they are getting the recommended quantity, and alter the feed combination as necessary.

Management of Bedding

Maintaining a clean and cosy environment for your hens requires routine bedding replacement. Make use of absorbent materials like pine pellets, wood shavings, or straw. To maintain the coop dry and odor-free, replace the bedding as needed. Proper bedding gives your hens a cosy place to sleep and helps regulate ammonia levels.

Health Surveillance

Checking hens for parasites like lice or mites should be part of routine health checks. Examine their skin, beaks, and feathers on a regular basis.

Keep an eye out for any odd bumps, lumps, or indications of distress. It is important to adhere to a veterinarian's recommended plan for regular deworming and vaccinations.

Cooperative Upkeep

Examining the chicken coop for wear and tear or damage is part of routine maintenance. Make that the walls, floors, and roof of the coop are in good shape.

To guarantee the safety and security of the coop, fix any broken or damaged pieces.

Gathering Eggs

Daily egg collecting is required from chickens that lay eggs in order to preserve egg quality and avoid egg breakage. Gently gather eggs and keep them somewhere dry and clean. Keep an eye on the quantity and quality of eggs produced, since variations may point to environmental stress or health problems.

Observing The Behaviour Of Chickens
Recognising Typical Behaviour

A variety of breed-typical behaviours are displayed by Polish chickens. Get comfortable with common actions like dust bathing, pecking, and scratching. By keeping an eye on these behaviours, one can see any deviations that can indicate environmental or health problems.

Recognising Stress Signs

Stress has an impact on the well-being and output of chickens. Excessive vocalisation, feather plucking, or aggressive behaviour are common indicators of stress. Recognise possible stressors and take immediate action to mitigate them, such as insufficient ventilation, crowded conditions, or abrupt changes in the surroundings.

Modifications in Behaviour

Keep an eye out for any behavioural changes that can point to a sickness or discomfort. For example, a sharp decline in activity or adjustments to eating or drinking patterns may indicate health problems before they become serious. Frequent behavioural observations assist in identifying issues early on and keeping them from getting worse.

Social Exchanges

Like other breeds, Polish chickens are arranged into social hierarchies and pecking orders. Check for bullying or hostility in the way the chickens interact with one another. Having enough room and resources reduces conflict and fosters a peaceful environment within the flock.

Enhancement of Environment

Give your hens access to feed, dust baths, and perches as environmental enrichment to keep them from getting bored. Enrichment promotes the flock's general wellbeing and aids in igniting natural behaviours.

Maintaining And Managing Records
Maintaining Records Is Essential

Maintaining accurate records is essential to managing chickens effectively. Maintaining thorough records enables you to monitor your flock's general health, production, and welfare. Keeping thorough records also makes it easier to spot trends and make wise management choices.

Kinds of Documents to Keep

Records pertaining to feeding regimens, water consumption, health observations, egg production, and cleaning procedures are all considered essential. Establish a system to capture these details every day and make sure the data is updated on a regular basis. To stay organised, use a log book or spreadsheet.

Medical Records

Keep track of every chicken's medical history, including immunisations, deworming procedures, and any ailments or injuries. By keeping note of these particulars, it becomes easier to monitor each bird's medical history and improve care management.

Measures of Performance

Monitor key performance indicators (KPIs) include growth rates, feed conversion ratios, and egg production rates. This information aids in evaluating your flock's productivity and pinpointing any areas in need of development. To assess trends and make necessary modifications, compare the measurements that are used today with the data from the past.

Examine and Interpret

Examine and evaluate your records on a regular basis to look for trends or problems. Keep an eye out for patterns in health issues, shifts in output, or behavioural modifications. Make educated judgements regarding flock management, feed modifications, and other hen care issues by using this knowledge.

Your Polish chickens' production and well-being may be guaranteed by according to these comprehensive instructions for daily care, regular chores, behaviour monitoring, and record-keeping. This will set the stage for a fruitful and satisfying experience with poultry farming.

CHAPTER EIGHT

Production And Management Of Eggs

Knowing The Cycles Of Egg Production

Polish hens are well-known for their unusual looks and kind disposition, but they also produce a lot of eggs. To maximise farm productivity and maintain flock health, you must comprehend the Polish chicken egg production cycle.

Like other breeds, Polish hens produce eggs in a cyclical manner. Hens usually start laying eggs when they are between five and six months old. The pullet phase is the name given to this first stage. The hen's reproductive system develops during this period, and she gradually begins to lay eggs. The size and frequency of the eggs stabilise as the hen's body adjusts, albeit the early eggs may be smaller and less numerous.

There are three phases to the egg production cycle: laying, peak production, and moulting. Regular egg production characterises the laying phase. Though this can vary, Polish hens often lay eggs every day or every other day. Hens lay the most eggs between the ages of 6 and 12 months, which is when they reach peak productivity. Following this peak, chickens will naturally go into a resting and rejuvenating cycle, which may result in a minor decline in the frequency of egg-laying.

Recognising how outside influences affect egg production is essential for regulating the process. Egg production cycles are greatly influenced by factors such as nutrition, lighting, and general health. To encourage egg production, make sure your Polish hens are fed a well-balanced diet high in calcium and protein. Furthermore, regular lighting that replicates daylight helps control their reproductive cycles, which results in more regular egg laying.

Gathering And Preserving Eggs

To preserve their quality and guarantee that they are safe for food, eggs must be collected and stored properly. In order to keep eggs from being dirty or damaged, they should be gathered at least twice a day. To reduce the chance of infection, handle eggs with clean, dry hands or gloves. Gently gather eggs to prevent breakage and maintain their cleanliness.

Eggs should be washed if needed after being gathered. But it's crucial to refrain from washing eggs unless they are obviously filthy. The natural protective layer may be removed by washing, which raises the possibility of bacterial contamination. Instead, use a soft brush or a dry cloth to gently brush out any debris.

Ensuring the freshness of eggs requires careful storage. Eggs should be kept cold and dry, preferably in a refrigerator at 45°F/7°C or less. To prevent eggs from acquiring strong flavours and odours from other foods, they should be stored in their original carton. The carton also helps keep eggs from drying out and losing moisture.

The pointy end of the eggs should be pointing down when being stored. This position lessens the chance of the yolk shattering and helps maintain it centred. Check eggs that have been stored frequently for indications of spoiling, such as an odd odour or strange discolouration.

The yolk of fresh eggs is firm and holds up nicely in the white, and they smell clean and neutral.

Handling Typical Egg Problems

Several typical egg problems can occur even under cautious control. Your Polish hens' health and the calibre of their eggs can be guaranteed if you are aware of these problems and know how to resolve them.

1. Broken Eggs

Eggs may crack as a result of improper handling, malnutrition, or unsuitable housing. When collecting eggs, handle them gently to reduce cracks, and make sure the chickens are fed a healthy diet that includes enough calcium to fortify their eggshells. Examine the diet and living circumstances of your chickens if you observe a high percentage of cracked eggs to rule out any possible problems.

2. Eggs with Soft Shells

Another prevalent issue is soft-shelled eggs, which are frequently the consequence of stress or a calcium deficit. To promote healthy eggshell creation, make sure your hens have access to a calcium supplement, like crushed oyster shells. Examine any environmental stressors that your flock may be experiencing, including as crowding or schedule changes, as these may also be having an impact on egg production.

3. Eggs with Two Yolks

When a hen releases two yolks in a single laying cycle, the result is double-yolk eggs. Double-yolk eggs are not dangerous,

although they may be a sign of stress or hormonal abnormalities. Although they are usually safe to consume, you should keep an eye on them to make sure they don't indicate any underlying health problems in your flock.

4. Gory Eggs

Injuries or infections to the hen's reproductive system can cause bloody eggs. Check your chickens for any symptoms of disease or injury if you observe bloody eggs, and get veterinary advice if needed. Preventing such problems can be achieved by making sure your hens live in a clean and secure environment.

5. Egg Attachment

When a hen can't pass an egg, it's called egg binding, and it can be a serious condition that needs to be treated right away. Lethargy, straining, and an enlarged abdomen are some of the symptoms. For treatment options, speak with a veterinarian right once if you suspect egg binding.

You can contribute to ensuring that your Polish chickens are healthy and productive and that their eggs are of the highest calibre by being aware of these aspects of egg production and management.

CHAPTER NINE

Overview Of Seasonal Factors

Depending on the season, Polish hens, who are distinguished by their colourful crests and amiable disposition, need different kinds of care. It is essential to comprehend these seasonal fluctuations if they are to continue being healthy and productive all year round. We'll examine how various seasons affect Polish chicken farming in this chapter and offer doable solutions for each.

Changing Up Care For Various Seasons
Spring: Growth and Rejuvenation

Polish hens go from the dormant winter months to a period of heightened activity and growth in the spring, a season of rejuvenation. It is important to maintain their nutritional needs and get them ready for the approaching egg-laying season during this time of year. Make sure they consume a well-balanced, high-protein diet to encourage the growth of their feathers and general health. Furthermore, look for any indications of mould or moisture in their living space, as these issues tend to become more common as the temperature rises.

Summertime: Keeping Cool and Hydrated

The summertime has unique concerns, especially with regard to heat stress. Because of their unusual feathering, Polish hens may be particularly sensitive to extreme temperatures. Provide plenty of shade and make sure their water supply is constantly cold and fresh to help reduce heat stress. To assist in reducing the temperature in their coop, you can also utilise fans or misting systems. It's crucial to keep an eye out for overheating symptoms in their behaviour, such as heavy panting or lethargic behaviour. You can assist them keep hydrated and sustain their energy levels by adding electrolyte supplements to their water.

Fall: Getting Ready for the Freeze

Polish poultry must progressively acclimatise to lower temperatures as the weather cools. Make sure their coop is insulated against the impending winter before starting to prepare it. Polish chickens benefit from a draft-free environment even though they are sturdy birds. By continuing to give them a healthy meal and making sure their water doesn't freeze, you can lower the chance of disease. It's also a good idea to look for any possible problems, such leaks or inadequate ventilation, with their living area.

Winter: How to Handle the Stress of the Cold

The worst season for Polish chickens can be the winter. They must be sheltered from inclement weather while yet having access to exercise opportunities. Make sure the coop is adequately aired and well-insulated to avoid moisture accumulation, which may cause respiratory problems. To keep them warm, give them extra bedding, and make sure their water supply is never frozen and is constantly available. During the shorter days, additional lighting can support the maintenance of their laying patterns and general health.

Handling The Stress Of Heat And Cold
Controlling Heat Stress

Several techniques are used to manage heat stress in order to keep Polish hens comfortable in hot weather. First, make sure there is lots of shade, either from man-made structures or natural elements like trees. If at all feasible, cool their coop using air conditioning or fans. It's imperative to make sure they have access to cold, clean water at all times; you may want to consider chilling their water bottles with ice cubes.

Treats that aid in cooling them down, like frozen fruits or veggies, are another option. Keep an eye out for indicators of heat stress, such as decreased egg production or increased panting, and intervene quickly if they materialise.

Handling Cold Stress

Make an effort to create a warm, dry atmosphere in order to control chilly tension. Make sure the coop is sealed against drafts and has adequate insulation

. If necessary, you can use heaters or heat lamps, but use caution around potential fire hazards. Add more bedding, like hay or straw, to the coop to aid in insulation.

To assist them maintain their body heat, give them extra nutrition and make sure their water isn't frozen on a regular basis. During the winter months, Polish hens might also benefit from a source of fat and protein, like broken corn or leftover kitchen trash.

Seasonal Wellness Advice
Springtime Wellness Advice

Check your Polish hens for indications of parasites in the spring, as they might become more active as the temperature rises. Check for mites or lice frequently, and take quick action to treat them.

To identify any problems early, make sure their immunisations are current and think about getting a thorough health checkup.

Summertime Wellness Advice

Throughout the summer, pay attention to their fluid intake and look out for any symptoms of heat-related ailments. Keep their coop clean on a regular basis to avoid the growth of parasites and dangerous bacteria. For natural parasite control, think about adding a dust bath area with diatomaceous earth.

Tips for Fall Health

Keep an eye out for early symptoms of respiratory problems or other ailments as the weather cools.

Keep an eye out for any indications of mould growth or moisture in their coop, as these conditions can worsen respiratory issues. As they get ready for winter, keep feeding them a healthy, balanced diet to maintain their wellbeing.

Wintertime Health Advice

During the winter, make sure their coop is well-insulated and draft-free to help prevent frostbite. Check their combs and feet frequently for indications of frostbite, and give them extra blankets to stay warm.

Throughout the winter, keep an eye on their water supply to keep it from freezing and give them extra nutrients to boost their immune system.

Understanding and taking into account these seasonal factors can help you keep your Polish hens happy, healthy, and productive all year long.

CHAPTER TEN

Promoting And Getting Rid Of Your Chickens

Recognising Demand In The Market

Marketing and selling your hens successfully depends on your ability to understand market demand. This entails investigating and evaluating consumer preferences, area needs, and current trends. Start by determining who your target market is, be it speciality markets, local customers, or eateries. There are specific expectations and needs for each sector. For example, while businesses may value regularity and mass availability, local customers may prefer fresh chickens raised organically.

Market research might yield insightful information. You can learn more about what customers want by conducting focus groups, surveys, and competitor analysis. Keep an eye out for shifts in consumer behaviour and seasonal patterns. For instance, during holidays or other special occasions, there may be a rise in demand for poultry. Furthermore, take into account regional laws and health trends that may have an impact on chicken consumption, such as the growing popularity of hormone- and antibiotic-free poultry.

Modify your manufacturing and marketing plans based on the data you have gathered from your investigation. Make sure the products you sell meet the needs of the market. This could be adding several varieties of chicken to your product line or providing speciality breeds that target specific consumers. You may effectively position your poultry business and build a devoted consumer base by remaining aware of market expectations.

determining prices and distribution channels

A successful chicken farming business depends on setting prices and selecting distribution channels. Pricing has to take into account the market rate, your production costs, and the calibre of your goods. Compute all related expenses, such as labour, feed, equipment, and overheads. Depending on your market positioning, take into account various pricing techniques including value-based pricing, cost-plus pricing, or competitive pricing.

A key component of reaching your target audience is selling channels. Local eateries, grocery stores, and farmers' markets are examples of traditional routes. Getting to know the managers of the grocery stores or local chefs will help you get frequent orders. Don't forget to take into account direct sales to customers via farm stands or internet channels. You may reach a wider consumer base and penetrate markets outside of your local area with online sales.

Investigate joint ventures with nearby companies or civic associations to boost exposure and boost revenue. For example, you can reach out to new consumers by partnering with regional food festivals or health fairs. Offering bulk discounts or putting in place a subscription model can also draw in repeat business and guarantee consistent income.

Review your pricing and sales tactics frequently in order to maximise profitability and adjust to changes in the market. You may efficiently market and sell your hens while realising the full potential of your company by picking your selling channels wisely and establishing prices that are competitive.

how to market your poultry enterprise

Marketing is crucial to drawing clients and creating a powerful brand identity for your chicken business. Start by creating a thorough marketing strategy that incorporates offline and internet tactics. Make a polished website that highlights your goods, tells visitors about your agricultural methods, and, if necessary, facilitates online buying.

Make use of social media channels to interact with prospective clients and provide information on your poultry company. Post information about your hens on a regular basis, such as recipes, customer testimonials, and behind-the-scenes photos of your farm. In order to reach a wider audience, social media advertising can also target particular demographics and geographic regions.

Building a network in your neighbourhood can help you become more visible. Engage in conversations with nearby businesses and prospective clients by going to community events, farmers' markets, and agricultural fairs. Distributing business cards, fliers, and other marketing materials in key places will help people learn more about your poultry goods.

To draw in and keep clients, think about providing loyalty programs, discounts, or promotions. Incentives for referrals or exclusive deals for first-time customers can promote word-of-mouth advertising. A solid reputation that is earned by high-quality goods and first-rate customer support is essential for both long-term business success and client happiness.

Recognising Demand In The Market
Examining Customer Preferences

Conducting customer research is the first step towards efficiently marketing and selling your chickens. It's critical to comprehend what prospective clients are looking for when customising your goods and services. To learn more about the sorts, flavours, and agricultural practices that local consumers want for chicken, do surveys or interviews with them. Observe shifts in preferences for organic or free-range products, as these may have an impact on purchasing choices.

To find more general trends, examine market research studies and industry reports. For example, the growing trend of health-conscious eating could increase demand for hormone- and antibiotic-free chickens. Following these trends enables you to modify your products and services to better suit the needs of your target market.

Examining Rival Approaches

Another crucial component of comprehending market demand is researching your competition. Determine any gaps or opportunities in the market by assessing the services provided by the other chicken farms in your vicinity. Evaluate their marketing tactics, pricing, and selection of products. You can use this study to identify your items' unique selling propositions and distinguish your company from the competition.

Think about what sets your poultry business apart, be it a particular breed, organic farming methods, or exceptional quality. Emphasise these unique selling points in your marketing campaigns to draw in clients who respect what you have to offer. You may increase your market share by strategically placing your products and filling up holes in the market.

Determining Prices And Distribution Channels

Pricing and Cost Calculations

Appropriately pricing your hens entails figuring out all related expenses and arriving at a price that equals your outlays plus the chickens' market value. Determine the total cost of production, taking into account labour, veterinary care, housing, and feed. Include overhead expenses like utilities and equipment upkeep. After you have a firm grasp on your overall costs, determine a pricing that will both guarantee profitability and keep you competitive in the market.

To appeal to diverse customer segments, think about utilising distinct price tactics. Offering premium prices for organic or speciality chickens, for example, can draw in affluent customers, whilst ordinary items with reasonable pricing can draw in budget-conscious shoppers

. To stay profitable, assess and modify your prices frequently based on production costs and market conditions.

Examining Different Selling Channels

Effectively reaching your target market depends on your choice of selling channels. Conventional avenues like restaurants, local food stores, and farmers' markets are excellent places to start. Developing connections with nearby companies can result in more visibility and regular orders.

To increase your reach, think about internet selling outlets as well. Establishing an online store enables clients to place orders straight with you, offering ease and possibly expanding your customer base. Social media sites and online markets can be useful tools for advertising and product sales.

Analyse the performance of every selling channel and modify your plan as necessary. For instance, you might wish to improve the functioning of your website or make an investment in digital marketing

if you see an increase in online sales. You can increase your sales chances by broadening your selling channels and frequently evaluating their effectiveness.

How To Market Your Poultry Enterprise

Formulating a Marketing Strategy

A well-thought-out marketing strategy is essential to successfully promote your chicken company. Establish your target market and unique selling propositions as part of your brand identity first. Craft a captivating brand narrative that appeals to consumers and emphasises the unique qualities of your company.

Create and carry out marketing plans for a variety of platforms, such as online and offline initiatives. Create a polished website and interact with your audience on social media. To keep customers informed and interested in your items, share promos, behind-the-scenes content, and updates on a frequent basis.

Making the Most of Community Involvement

Getting involved in the community can greatly increase the awareness of your chicken company. Engage in local gatherings, farmers' markets, and agricultural exhibits to establish connections with prospective clients and exhibit your merchandise. Developing connections with businesses and organisations in the area might help you become more visible and credible.

To inform the community about your farming methods and goods, think about holding workshops or farm tours. These occasions can increase client loyalty and present chances for in-person sales. Interacting with the community improves client relations and establishes a favourable reputation for your company in addition to promoting it.

CHAPTER ELEVEN

Faqs & Troubleshooting

Solving Typical Problems In Polish Chicken Farming

1. Problems with Feathers

In Polish chicken farming, plucking of feathers and irregular growth of feathers can pose serious problems. Stress, malnutrition, and overcrowding are common causes of feather plucking. Make sure the hens have enough room to roam about and eat a well-balanced diet full of vital nutrients in order to solve this. Furthermore, look for parasites that could harm feathers, such as lice or mites.

2. Low Level of Egg Production

Numerous variables, including as inadequate illumination, poor diet, or stress, might contribute to low egg production. Make sure the chickens get 14–16 hours of light a day to encourage the production of eggs. To produce eggs as much as possible, a balanced diet rich in calcium and protein is essential. Any underlying medical conditions that may be impacting egg laying can be found with regular checkups.

3. Health Issues

Similar to other breeds, Polish chickens are prone to respiratory illnesses, coccidiosis, and Marek's disease. A clean environment, enough ventilation in the coop, and routine vaccines can all aid in the prevention of these illnesses. For a diagnosis and course of therapy, speak with a veterinarian if you observe any symptoms of sickness, such as lethargy or strange behaviour.

4. Conduct Problems

Chickens that engage in bullying or aggressive behaviour can cause discord in the flock. To avoid rivalry, make sure there is adequate room and supplies, such as feeders and waterers. Stress and violence can be decreased by setting up hiding places and offering enrichment activities.

5. Environmental Elements

The health and welfare of Polish hens might be impacted by harsh weather. Make sure they have enough shelter to withstand the cold and that there is enough airflow to keep them from becoming too hot. To make your hens feel at ease, keep an eye on the temperature and humidity inside the coop.

Commonly Asked Questions

1. What distinguishes Polish chickens from other types of chickens?

Polish chickens are distinguished by the characteristic feather crests on their heads, which give them a distinctive look. They are a decorative breed that is frequently maintained more for show than for usefulness. Additionally well-known for their friendliness and curiosity are Polish hens.

2. Taking care of Polish chickens: how do I?

A balanced meal, clean water, and a safe coop are all part of proper care. Polish hens need a diet high in calcium and protein, as well as routine check-ups and a tidy living space. To avoid any problems with vision or head health, it's critical to maintain their distinctive feathered crests.

3. How can I determine the health status of my Polish hens?

Polish hens that are ill may exhibit altered behaviour, decreased appetite, unusual faeces, and outward displays of distress. Keep a close eye on your chickens and get advice from a veterinarian if you observe any signs of disease.

4. Is it possible to keep Polish chickens with other breeds?

It's possible to keep Polish hens with other breeds, but you should keep an eye on how they get along. They could be more susceptible to bullying from more aggressive breeds because of their distinctive appearance. Make sure the flock has adequate room and supplies to minimise any possible confrontations.

5. Which housing techniques work best for Polish chickens?

Polish hens need a coop that is well-ventilated and has enough room for them to walk about in. Make sure the coop is protected from harsh weather and is safe from predators. Maintaining a clean and well-maintained environment is crucial for the health of your chickens.

Sources Of Additional Support

1. Farming Associations for Poultry

Joining local poultry clubs or poultry farming associations like the American Poultry Association can offer helpful resources, networking opportunities, and professional guidance on managing and caring for Polish chickens.

2. Internet Communities and Forums

Social media organisations and online forums devoted to chicken farming can answer particular queries, share experiences, and offer peer support. Communities of people who are passionate about chickens may be found on websites like Poultry Keeper and Backyard Chickens.

3. Veterinarian Services

For assistance in identifying and managing health problems, speak with a veterinarian who specialises in poultry. Seek for vets with experience caring for poultry if you want your

Polish chickens to receive the best care possible.

4. Books and Handbooks

Books about raising poultry and specialised care manuals for Polish chickens can provide detailed knowledge and helpful advice. Take into account publications like Gail Damerow's "The Chicken Health Handbook" and other guides on caring for chickens.

5. Workshops & Seminars for Education

You can improve your knowledge and abilities about poultry farming by taking part in workshops, seminars, or online courses. Seek for gatherings organised by organisations engaged in poultry farming or agricultural extension.

Typical Problems And Their Fixes
1. Health Concerns

Challenge: Respiratory diseases and parasites are among the health issues that Polish hens are susceptible to.

Solution: Numerous health problems can be avoided with routine physicals and immunisations. Keep your coop clean, and feed your hens well to boost their immune systems. See a poultry veterinarian for advice if health issues emerge.

1. Taking Care of Feathers

Challenge: Polish chickens' characteristic crests can get tangled or obstructed, which might cause health issues.

Solution: Maintaining proper grooming is crucial. Make sure there is nothing in their surroundings that could catch in their feathers, and carefully trim the feathers surrounding their crests. Maintaining the health of feathers also involves giving them enough room and minimising stress.

1. Production of Eggs

Challenge: A variety of factors, including age, food, and environmental circumstances, can make ensuring consistent egg production challenging.

Solution: Give laying hens a premium meal that satisfies their dietary requirements. To replicate the cycles of natural daylight, keep lighting constant. To promote continuous egg production, control stressors within the flock.

1. Space Administration

Challenge: If raised in overcrowded conditions, Polish hens may exhibit signs of stress and hostility.

The solution is to make sure that the coop and run are large enough for the quantity of hens you own. Use tactics like numerous feeders and waterers to lessen competition for resources, and offer enrichment activities to keep them occupied.

1. Environmental Elements

Challenge: The health of Polish chickens might be impacted by high temperatures and inadequate ventilation.

The solution is to build your coop with enough ventilation and insulation to guard against temperature fluctuations. In order to keep the atmosphere comfortable, use heaters or fans as needed. Keep an eye on the weather and make necessary adjustments depending on observations.

conclusion

Commonly Asked Questions

1. How frequently should I look for health problems in my Polish chickens?

At least one regular checkup should be carried out each week. Keep an eye out for symptoms of disease, including as behavioural changes, droppings, and the state of the feathers. Effective health problem management requires early detection.

2. What kind of food should Polish chickens eat?

Polish hens need a well-balanced diet that includes grains, veggies, and even treats in addition to premium poultry feed. Make sure the meal has enough calcium and protein, especially for laying chickens.

3. How can I stop my Polish chickens from plucking their feathers?

By giving birds adequate room, lowering stress levels, and making sure they consume a healthy diet, feather plucking can be decreased. To keep them interested and prevent boredom, think about introducing perches and other environmental enrichments.

4. If my Polish hens aren't producing eggs, what should I do?

Look into possible causes include bad nutrition, insufficient illumination, or medical conditions. Make sure they eat a balanced diet and have adequate lighting. Seek additional guidance from a poultry specialist if issues continue.

5. What is the proper way to care for Polish hens in severe weather?

Make sure there is enough shelter to fend off the heat, rain, and snow. Make sure the coop has adequate ventilation in the summer to avoid overheating and insulation in the winter to keep the cold out. To ensure your hens are comfortable, keep an eye on the weather and make any necessary alterations.

Sources of Additional Support

1. Websites Concerning Poultry Health

Important information about chicken health, illness prevention, and general care advice can be found on websites like the Poultry Extension at several agricultural universities.

2. Journals & Magazines for Poultry

You can stay up to date on the newest findings, industry trends, and best practices in chicken farming by subscribing to poultry periodicals or journals like "Poultry World."

3. Services for Local Agricultural Extension

Local agricultural extension services provide professional guidance, courses, and materials catered to the unique requirements and circumstances of your area.

4. Books about Poultry Farming

Books like Gail Damerow's "Storey's Guide to Raising Chickens" provide thorough instructions on caring for chicken, including advanced management strategies and solutions for frequent issues.

5. Tutorials & Courses Online

Courses on chicken farming are available online on sites like Coursera and Udemy. They cover a range of topics from basic care to sophisticated management strategies.

..

THE END

THE END